GAMPHACE

An American Story

The End of Prohibitionon on Federal Cannabis

"Excellence is doing a common
thing in a uncommon way."
- Booker T. Washington

ACKNOWLEDGEMENT

We would like to thank Robert-H El-Bey and Elder John for their head strong mentality and the absence of fear carried within their hearts. In a world full of individuals who will not stand for themselves or others. Their work has positively affected masses even myself and family.

-From the Members of Gamphace.

INTRO

The reason I was led to publish this book and the GAMPHACE SERIES is to express the truth. To ensure this story is told the right way. The factual way. Also to give the people an option that they didnt know existed due to corporate control and cover ups for gains in taxation without representation.

Its my intention that this message transpires beyond these pages and on to a larger stage to reach as many people as possible.

WHAT IS GAMPHACE

GAMPHACE itself is many things. A FEDERALLY APPROVED CANNABIS COMPANY that works in the research and development since 2004, with Federal Schedule 1 (Marijuana) Exempt Industrial Hemp. In January of 2020 there is no company that can operate in the same legal/lawful capacity as GAMPHACE /GROWING PHARMS. Mainly due to the Federal exemption they have under CSA 885a1 { https://www.deadiversion.usdoj.gov/21cfr/21usc/885.htm} in collaboration with FEDERAL SCHEDULE 1 MARIJUANA-EXEMPT-INDUSTRIAL HEMP OF NEW JERSEY organized exclusively to contract with entities such as corporations, non-Governmental, Governmental, Public and Private Banking Institutions, National and International Agencies, The original, Ab-Original and indigenous Peoples of the Globe. Adopting the Constitution (of) and (for) the United States and the Organization of the Americas States [OAS]. Adopting the laws of the United Nation Rights of Indigenous People under the International Laws and holding officers to their oath of office, and accountability in accordance of FEDERAL CIVIL RIGHTS STATUES; Title 18, U.S.C. sections 241,242, and 245(1)(b). I.R.S. Department of the Treasury registration removes members from a suspected class. C.S.A. 822(c)(1)(2)(https://www.deadiversion.usdoj.gov/21cfr/21usc/822.htm) allows for possession of controlled substance (MARIJUANA), warehouse and carriage. I.R.S. Department of the Treasury Federal I.D. #'s (20-********) allows the manufacture of a Schedule 1 controlled substance (Marijuana). Private business entity and business (members) are exempt from all statues of alleged criminal activity by the closed regulatory system enforced by the Drug Enforcement Agency (D.E.A). The Business has a closed distribution system that allows product (INDUSTRIALHEMP-CANNABIS-(META-CENSATION©™) for sale to members who are ultimate users; (there is no conveyance to a third party.) For the record. Business to apply a Sales of Tangible Personal Property for use directly and exclusively in experimental research and development in the laboratory sense. (tax). Research and development of [INDUSTRIAL-HEMP-CANNABISFLOWER] including "META-CENSATION©™" Carbon Sequestration.The only other business model even close to what they have is the University of Ole Miss but they are heavily restricted and regulated by the DEA and do not have the ability to Resale as GP does. Of course, you are reading this trying to figure out how it's even possible or saying this can't be real even though you just looked at the D.O.J (Department of Justice) links. Don't worry we will get to the how soon.

Back in 2003, the Elder Robert established Gamphace as a music company, mainly dealing in hip hop. Guns And Money ,Power, Hoes ASS ETERNALLY for 2 artist. Originally the acronym you see was made to express the active and fast pace lifestyle Robert was living at the time. By 2006, Robert seen it fit to make a change not only for himself but the brand GAMPHACE as well. One day he went home and looked at his son as he played his PlayStation PSP. Overcome with a strong sensxe of purpose. His eyes swelling with tears. His hand went into auto pilot on the paper, and letters went from negative to positive.

G-GAMEPLAN

A-ATTITUDE

M-MOTIVATION

P-PRIDE

H-HUNGER

A-AMBITION

C-CONFIDENCE

E-EXCELLENCE

Robert said to himself, "I have something here for everyone and it will last forever. How do I get it to them?" His brother instantly came to mind. He had a moving company and Robert had been in and out of the relocation business since 1993. So he gave him a call and told him "I'm getting on the truck with you".

He Asked, "You sure".

"Yes", Robert responded

Over the next two years Robert and his brother went state to state every day for years relocating people lives. Rob would take his own money he earned from work and buy 100 T-shirts at a time made with GAMPHACE GP logo on the front and the meaning of GAMPHACE on the back. Roberts goal was to get at least 5 GAMPHACE shirts in every State.

We The People...
united together for
Hemp Cannabis
Educational
Awareness

Original GAMPHACE Shirts

In the process relocating and meeting new people would ask how GAMPHACE was spelled. Everyone would say Game Face and he would spell it expressing that it was created to spread a positive message across the world. One day on the road his brother said, "Rob I don't get it I give you money and you go buy shirts to give away for free." For Rob it was about the message not the money. Hes always been able to see ahead, even to this day. With PRIDE people all over the states represented GAMPHACE . Due to the level of marketing and branding Robert put forth, Elder John reached out to Rob and told him, "Gamphace words of wisdom helped me get through the day on Facebook. You need to help me and come out to the farm" said John.

A trip came up for him to go to Cali on the moving truck . Rob was requested to babysit a new driver so it was not something he was looking forward to but it was a way for him to get to the west coast. He got dropped off in L.A. where he found a ride share off craigslist and went form L.A. to San Francisco. The driver was from Chile, Emanuel was his name . Emanuel asked Robert in his accent his prespective on medical marijuana. Rob responded "Medicine!?" He said "...fire that shit up."' The bud was so strong after smoking shortly after Robert passed out in the car fast asleep. Over an hour had passed and he realized he was high as hell and fell asleep in a car with complete strangers. So he peaked up and said, "Alright, if the waters on the left I'm good. If it's on the right we going south." He sat up from the right side of the car and looked to the left he seen only ocean . He felt a sense of relief. The driver saw him sit up and said " Aaayyyeee you awake? " Robert asked what was that we was smoking? Emanuel called it "Gods bud". After 5 hours in the car they finally got to the destination and as a parting gift, Robert gave them a GAMPHACE T-shirt.

Robert caught a train to Oakland. There he and John met for dinner. They went up to what he referenced as HEAVEN, and to the farm from there. As soon as Rob hit the fence line his 1st encounter was with Raider, a pitch black dog german shepherd. Raider was well trained and would patrol and protect the farm. John and Raider walked Robert around the property showing him different parts of the house. It was a large engraved door. "Beautiful" is what Rob said as he glanced at it. John said "Come on in." He swung the large door open and to any average individual they may have seen chaos. Rob took a step closer and seen it was nothing but legal document and case work with D.E.A. stamps on them and Federal documents from the White House. Robert was in amazement. John said, "Well its some weed over there, smoke as much as you want." Robert said to himself this really is Heaven. An hour went by John comes back and ask him "Well what do you think?"

Robert said " John this is AMAZING."

" Yeah I know so what do you think how much is it going to cost you to help me." John asked. In John's mind Robert would come to work for him, but Robert is not really the employee type of guy. He is a business owner. So Robert told him " $100,000,000 ". John stepped up to Robert in his face and said "WHO IN THE FUCK DO YOU THINK YOU ARE COMING IN TO MY HOUSE TALKING ABOUT 100 FUCKING MILLION DOLLARS."

Robert calmly responded, " If you calm down and step back I can get a cab and get a hotel. I have a family and a business to run home to." John stepped back and said "Well 100 Million is not even 1% of the market Robert so, Alright."

It was a test! Could you blame John after having the D.E.A illegally raid 400+ plants and genetics, even stains from the music artist Carlos Santana. Having it all covered up and given back to you. Only to shut you up so there is no legal action. Knowing you are the only one with a Federal Exemption for Cannabis with no regulations. You for sure will integrity check those you bring onto your team. You would be foolish not to.

What was initially just supposed to be about a week stay ended up being a month and a half of Robert and John building together. Growing together and looking for ways to expand the company and information about the end of prohibition on federal cannabis.

When is was time for Robert to leave John asked a favor of Robert, which was to stay out of the Medical Marijuana industry. Only dealing with the Industrial hemp. After about 75 days John granted Robert membership. While Robert prepared to leave for

Buffualo, John shared that he grew some Violator Kush. It was not fully dry but he wanted him to have some to take back home.

Rob was getting settled back in Jersey and was introduced to his 1st bud connect in town by his current Empress, when he got to Jersey. He was a Jamaican named Peter. This particular day Rob, accompanied by Peter, was driving and smoking the Violator Kush John gave him. For the first time, he had stuttered. The words META-SENSATION. Peter asked what the newly invented word meant. Rob responded "I don't know but I'm not smoking no more Marijuana. From now on its Meta-Sensation" (Also to highlight Robert was aware of the legal codes and statutes bound to the word MARIJUANA which is copyrighted by the United States and controlled and regulated by them in a research and development sense utilizing people as test subjects and tax generators. Also leading to the use of a better fitting term.)

(Meta-Sensation is a copyrighted trademarked term by GAMPHACE it is the byproduct from the biomass and the meta-physical feeling you get from partaking in the plant i.e the byproduct from the biomass. Also known as Cannabis and commonly referred to as weed.)

After about a month Robert got a call from John. "New Jerseys all yours, just file it in your county,' Robert did just that! The state gave the company ST3 forms for resale of the companies property, meaning cannabis . Also ST4 forms for Exempt Use for the Ultimate users.(Ultimate User is a person who has lawfully obtained and possess a controlled substance for his own use or for the use of a family member of his household ; for an animal owned by him or someone in the household). The documents stated, for in state and out of state property. Robert instantly knew he need a team. From there GAMPHACE became a network with members. Robert being #1 in the Federal Sch 1 Marijuana Exempt industrial Hemp.

THE 3 CHIEFS

Elder John

The starter of it all, Elder John , based out of California. He is reserved at times and the complete opposite at others. Mind his business type of guy because it makes cents. For that reason information such as his full name will remain confidential for his safety and the safety of the network and other tribal members. Its also time when you may catch John flying up a 30 mph road doing 90 screaming "Who in the fuck can touch me! They know who I am out here." He was not lying either. John and his father lived on on old war land. Back in 1925 the U.S. used the land John bought to live on to cultivate marijuan seeds aka hemp seeds to feed messenger pigions. Making that Federal land as u can see in the next inage. Our Elder John was granted the exemption first. Based on a court case due to an illegal raid on legal cannabis by the DEA (Drug Enforcement Agency) . John being the man he was refused to take such unlawful treatment. It turned into a huge lawsuit against the federal Government. It got to the point where John was so furious that he wanted to make it public and take it to trial. John traveled all the way from Oakland, California to Washington, DC to put in his complaint and while he was there he got stuck in an elevator with a gentleman by the name of Reggie Love “BODY MAN”. Former President of the United States Incorporated Barack Obama Personal Aid. Reggie Love received the documents from John and passed them down to the Obama Administration. In return John received a letter from the White House. That informed him that due to the separation of powers the , White House would have loved to help , but he had to take it to the judicial system. That’s exactly what he did next.

The D.O.J. was not going to allow the public to be aware of such a screw up. Also allowing the option for more people to come back and bring charges up against them for the same thing. So to stop that, a special counsel from Washington, DC sent by the Bush Administration was sent to John's case (about a year had gone by at this time). That special counsel approached John and started poking John in the chest telling him he would not be taking this case to trail. He had came to inform John that the government prosecutor deliberated and informed the Judge that this case is to be dismissed and he is granted the ability to cultivate on the Federal level. John then ask "...and the business?"

"Yes the business as well." The Judge even spoke it in disbelief at the time. Due to this ruling all property that was originally taken by the D.E.A. was forced to be returned, as well as what they classify as MARIJUANA.

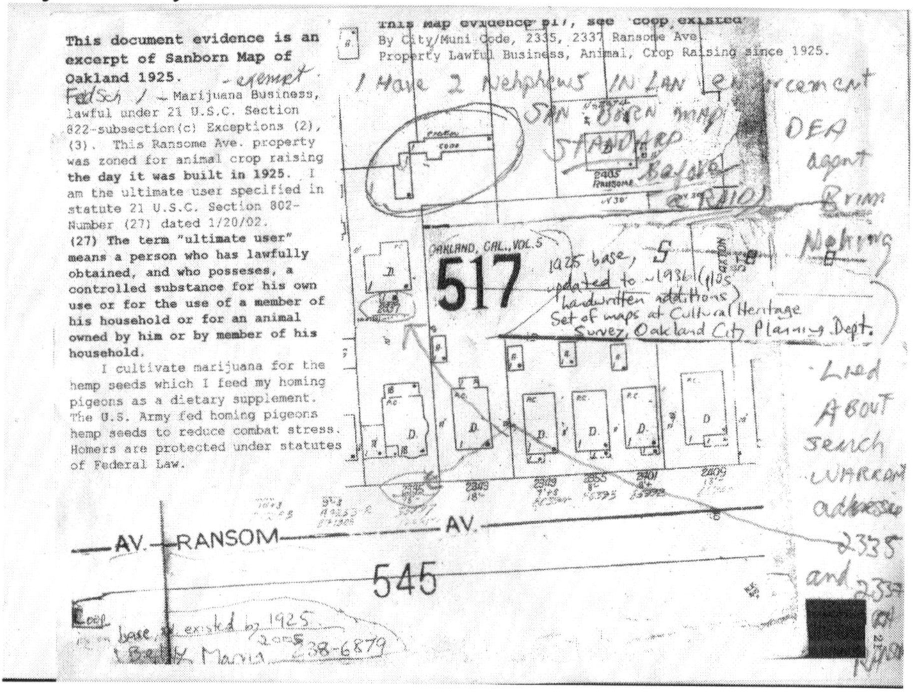

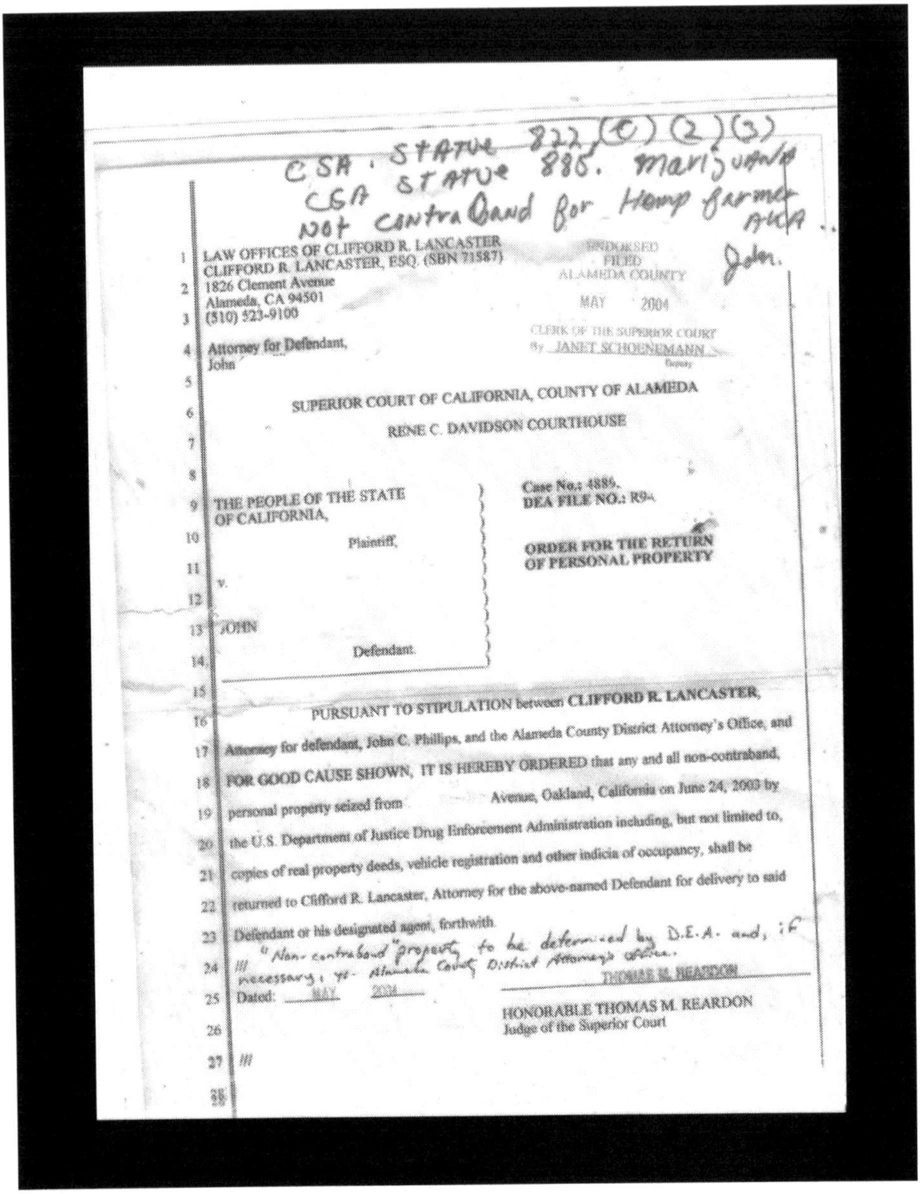

CSA : STATUE 822 (C) (2) (3)
CSA STATUE 885. marijuana
Not contraband for Hemp farmer AKA ..
John.

LAW OFFICES OF CLIFFORD R. LANCASTER
CLIFFORD R. LANCASTER, ESQ. (SBN 71587)
1826 Clement Avenue
Alameda, CA 94501
(510) 523-9100

Attorney for Defendant,
John

ENDORSED
FILED
ALAMEDA COUNTY

MAY 2004

CLERK OF THE SUPERIOR COURT
By JANET SCHOENEMANN
Deputy

SUPERIOR COURT OF CALIFORNIA, COUNTY OF ALAMEDA

RENE C. DAVIDSON COURTHOUSE

THE PEOPLE OF THE STATE OF CALIFORNIA,
Plaintiff,

v.

JOHN
Defendant.

Case No.: 4889,
DEA FILE NO.: R9

ORDER FOR THE RETURN OF PERSONAL PROPERTY

PURSUANT TO STIPULATION between CLIFFORD R. LANCASTER, Attorney for defendant, John C. Phillips, and the Alameda County District Attorney's Office, and FOR GOOD CAUSE SHOWN, IT IS HEREBY ORDERED that any and all non-contraband, personal property seized from Avenue, Oakland, California on June 24, 2003 by the U.S. Department of Justice Drug Enforcement Administration including, but not limited to, copies of real property deeds, vehicle registration and other indicia of occupancy, shall be returned to Clifford R. Lancaster, Attorney for the above-named Defendant for delivery to said Defendant or his designated agent, forthwith.

"Non-contraband" property to be determined by D.E.A. and, if necessary, the Alameda County District Attorney's office.

///

Dated: MAY 2004

THOMAS M. REARDON

HONORABLE THOMAS M. REARDON
Judge of the Superior Court

///

This is the start of the CSA 885a1 Federal Schedule 1 Marijuana Exempt Industrial Hemp network. On this day history was made towards the ending prohibition on cannabis silently.

U.S. medical-marijuana shops struggle to find bankers

By Brett Wolf, Thomson Reuters Accelus June 18, 2011

Non Believers------>>

Breaking barrires all the way back in 2011 just by being one of the 1st to open a legal / lawful bank account for cannabis. Due to the exemption and being removed from criminal suspected class by the D.E.A. and IRS. This was greter then just getting high on a plant . This was BUSINE$$.

Robert-H El-Bey

An American Indian born in Boston Massachusetts was introduced to John shortly after all this happened. Being a music producer/engineer also business owner of relocation company that specializes in moving people lives. After the elder John granted Robert his exemption and the ability to franchise out in multiple states and internationally. Robert is reserved but for years was on the front lines pushing for the government to express the truth about the end of federal cannabis prohibition. As you can see him in the picture with former Governor Chris Christie, holding the Federal Schedule 1 (Marijuana) Exempt Industrial hemp certificate and him showing presence at a political rally. On the following page we have Mike Mclntyre North Carolina State Representative and Jovan Roseboro. Jovan decied to embark on a spirital/motivational jorney to D.C from his home ON FOOT from Fayetteville , North Carloina. Moratorium on Selfless Walk. Robert got wind of Jovan would be walking through his hood. Robert was so moved by Jovan message. That he was willing to partake in a noble cause to stimulate change he reached out to Jovan friend and told her he had a gift for Jovan when he got to his town. She said “Really he would love that” Rob then followed up by saying “...and im going to walk the rest of the way to DC with him.” She was in disbelief. Even still she made the call to Jovan to inform him of the good news. Still two days away from DC they met and stopped and did a video upload in Stanford VA titled "1st person to join the walk Robert H Moore Jr' www.Gamphace.com. Jovan introduces Robert and explains how thankful he is just to have someone on the jorney with him. So with Jovan travling alone and with limited funds at the time Robert offers to put him up in a hotel room for the night. They wake and complete the walk to DC. There Jovon met with state reps and other officals wearing a Gamphace shirt in the White House. As you will see in the pictures Robert is no stranger to poltical interaction or standing for something he knows is right , not believes.

We the People

The County of Passaic

State Of New Jersey

The Passaic County Clerk
certifies the following
as a registered business
in Passaic County

FED SCH I MARIJUANA EXEMPT
INDUSTRIAL HEMP OF NEW JERSEY

Kristin M. Corrado, Esq., Passaic County Clerk

SEPTEMBER 11, 2012
File

Governor Chris Christie hosted a Town Hall Meeting at St. Luke Baptist Church in Paterson Tuesday with over 700 people in attendance.

Robert H. Moore Jr., Owner of GAMPHACE� a hemp business, asks Governor Chris Christie a question during Tuesday's Town Hall Meeting at St. Luke Baptist Church in Paterson.

1000 Watts
magazine
a bright idea for patients by patients
1000WattsMagazine.com
Interview with Robert H Moore Jr.
INSIDE THIS ISSUE
CANNAFRESH
PEAR SALAD
FOR
YEARS BY OUR
CANNABIS & CANCER:

Robert has even won court cases for cannabis himself due to the exemption and his own due diligence in his studies as he represented himself every time. It was for sure not always an easy experience when Robert had to deal with the Judicial System. In his mind this was the way to get it to trial to make it known that the exemption existed and the prohibition on cannabis ended. Through his experience that was not the case, all he came to know was moments where he was forced to educated judges and attorneys on something they didn't even know existed leading to case dismissal. They usually end up saying they no longer wish to file charges angst the defendant. Here are a few documents on dismissed cases. Some that Robert did not appear in court and the cases were still dismissed.

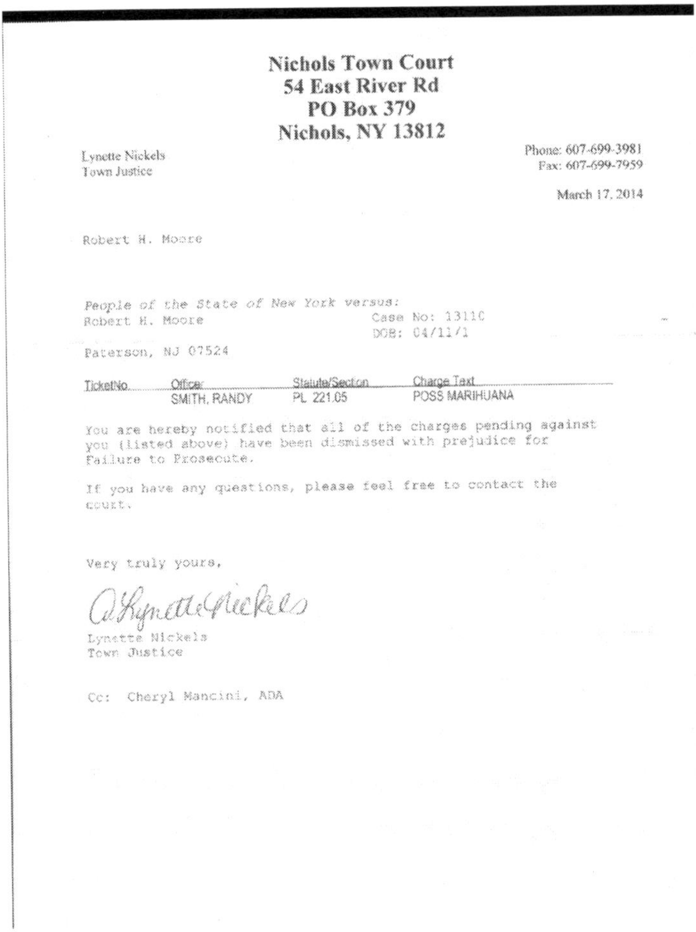

Nichols Town Court
54 East River Rd
PO Box 379
Nichols, NY 13812

Lynette Nickels
Town Justice

Phone: 607-699-3981
Fax: 607-699-7959

March 17, 2014

Robert H. Moore

People of the State of New York versus:
Robert H. Moore
Case No: 13110
DOB: 04/11/1

Paterson, NJ 07524

TicketNo	Officer	Statute/Section	Charge Text
	SMITH, RANDY	PL 221.05	POSS MARIHUANA

You are hereby notified that all of the charges pending against you (listed above) have been dismissed with prejudice for Failure to Prosecute.

If you have any questions, please feel free to contact the court.

Very truly yours,

Lynette Nickels

Lynette Nickels
Town Justice

Cc: Cheryl Mancini, ADA

STATE OF NEW YORK : COUNTY OF TIOGA
TOWN OF NICHOLS JUSTICE COURT

THE PEOPLE OF THE STATE OF NEW YORK,

- against -

ROBERT H. MOORE, JR.,

Defendant.

READINESS STATEMENT
& CPL 710.30 NOTICE

SIR:

PLEASE TAKE NOTICE that the People in the above-entitled action are in all respects ready and able to proceed to trial.

PLEASE TAKE FURTHER NOTICE, pursuant to the provisions of § 710.30 of the Criminal Procedure Law, the People intend to:

(   ) offer a confession or statement in evidence upon the trial of the defendant in the above-entitled action.

PLEASE TAKE FURTHER NOTICE that if the defendant intends to attack such confession or admission as involuntary, timely motion must be made to suppress such confession or admission from use as evidence.

Dated at Owego, New York
December 3, 2013

Cheryl A. Mancini
Tioga County Assistant District Attorney
PO Box 300
Owego NY 13827
Phone #607-687-8650

TO: Town of Nichols Justice Court

✓Robert H. Moore, Jr.
[redacted]
Patterson, NJ 07524

Kirk O. Martin
District Attorney

20 Court Street
P.O. Box 300
Owego, NY 13827
Tel. (607) 687-8650
Fax (607) 687-1614

Eric H. Garteman
First Assistant District Attorney

Cheryl A. Mancini
Assistant District Attorney

Palmer J. Pelella
Assistant District Attorney

OFFICE OF THE DISTRICT ATTORNEY

COUNTY OF TIOGA

March 7, 2014

Hon. Lynette Nickels
Town of Nichols Court
54 E. River Road
Nichols, New York 13812

RE: **People v. Robert Moore – Unlawful Possession of Marihuana**

Dear Judge Nickels:

The above-named defendant has a charge pending in your Court and was set for trial on the February court date. At that time, the People had numerous witnesses from the New York State Police and the New York State Police Crime Lab present for trial. However, the defendant did not appear and the matter was set for trial once again on March 10, 2014. Given the nature of the charge (a violation) and the expense that Tioga County has already put into prosecuting this charge, the People are at this time declining to prosecute the matter.

Thank you for your assistance in this matter. Please do not hesitate to contact me if you have any questions.

Respectfully submitted,

Cheryl Mancini
Assistant District Attorney

CAM/
Cc: Defendant

Nichols Town Court
54 East River Rd
PO Box 379
Nichols, NY 13812

Lynette Nickels
Town Justice

Phone: 607-699-3981
Fax: 607-699-7959

December 9, 2013

Robert H. Moore

People of the State of New York versus:
Robert H. Moore
Case No: 13110
DOB: 04/11,

TicketNo.	Officer	Statute/Section	Charge Text
	SMITH, RANDY	PL 221.05	POSS MARIHUANA

Trial date: 01/13/2014 Time: 04:00PM

This Court has accepted your not guilty plea. You have been scheduled for trial as shown above. If you plan to be represented by an attorney he or she should accompany you and be ready for trial as scheduled. Notices sent to attorneys are **NOT** copied to defendants. **No adjournments will be considered within 1-week of the trial date.**

If you wish to discuss a possible disposition with the District Attorney's Office prior to the above court date, you may contact: DA Irene Graven or ADA Cheryl Mancini at PO Box 300, Owego, NY 13827, (607)687-8650 or FAX (607)687-1614.

IMPORTANT: YOU MUST CONTACT THE COURT AT LEAST ONE WEEK PRIOR TO YOUR TRIAL DATE, IF YOU WILL NOT BE APPEARING AT TRIAL! THIS IS FOR COURT SCHEDULING PURPOSES.

A. Lynette Nickels
Lynette Nickels
Town Justice

Its was around 2014 when Robert met another cannabis advocate who is for sure worth the honorable mention, pioneer in the spheres of industrial uses for hemp and medical Cannabinoid based products with over 10 years of ground-breaking biotechnology research and exploration, Dr. Brian Gillis has developed unique extraction methods and complex products such as oils, tinctures, salves and vaporizer gels. In June, 2014, he was awarded an Honorary Doctorate of Divinity for outstanding service to the community in the field of biotechnology, and the healing of many through the use of the products he developed and shared. His research and development (R&D) corporate venture, "Canna – Dynamics," also known as "Canna-Dyne," is expanding to provide quality cannabinoid based products, medicines, delivery systems strain research, network and media platforms, and more.

He came on board with the team and created top tier genetics. Along with his information and that of the rest of the federal farmers have be able to make magnums ground breaking discoveries all in the federal private research and development sector.

DJ RedHorse Jones Bey

Also known as the Prince of Hemp from Detroit Michigan. Working as a public figure and a financial advisor for A.T.U.M. an Indian tribe. Was introduced to Robert's work by a mutual partner Sims. DJ and Rob initial introduction was on the phone,14 years down the line in 2018. You can imagine how someone with something with such value would get tired of people always trying to tell him what to do or how to run GAMPHACE or how to utilize their exemption. Even still when Robert speaks he is calm and direct. Always open for reason and growth. So Sims who introduced DJ to Rob. Mainly because he seen benefits in utilizing DJ to execute some of his own personal plans. He requested that DJ become a representative for GAMPHACE. Now the way Sims expressed it to DJ was that he was given a Federal License for Cannabis. At this time DJ had already been growing and making vegan edible for going on 8 years in Detroit. So he was willing and able to assist in the growth of the company. In his mind, thinking this is some exclusive shit. Being young and eager to make something shake. He set up a meeting for the other new reps to meet at the Michigan Hash Bash In Ann Arbor. He got right to work going to events. Sharing the information about the company. DJ learned quickly how ignorant people can be. Coming in contact with multiple growers and dispensary owners all believing there is no Federal grow operation. DJ realized fast Michigan was not the right place to rollout something this major. He said to himself "They think to small here and thats why poverty is so high. You know better follow the money". After extensive research in marketing, cannabis, different state operations about a month later Sims got upset with DJ for marketing the company online. In DJs mind this is how he has built out multiple companies. Make a website, create content and share it. He told DJ he was moving to fast but refused to be clear in the error DJ made. Now coming from a state where cannabis is legal for medical use already. He knows it's the exact opposite. He said "We are behind we don't need to slow down we should be going faster" and that didn't sit well with Sims. Sims had the audacity to tell DJ he felt untrusted and he should just believe in him. Coming from Detroit, that didn't fly. He told Sims "I don't believe shit. I deal in what I know." Sims got quite. DJ just knew this was going to be an issue for him and he said, " These geriatrics trying to hold me up."

He took it upon himself to reach out to Robert without any 3rd parties. DJ messages Robert on FACEBOOK and said "Peace Rob how are you this is DJ we spoke on the phone a while back. I'm interested in doing somethings things with the company but feel like I'm hitting roadblocks and would like to speak if possible."

Rob message back,"sure I will be in Michigan around this time." Rob gave DJ a call letting him know he was onthe way. DJ jumped in his truck. An F150 Super duty on a lift kit with 36-inch tires with rims pulled out still on the phone with Robert. DJ pulls out on the main street and sees a white box truck and says to Robert " Is this you?" Robert says "Yes in the white box truck" Dj then tell him to follow him and they park the trucks on the street behind Dj Family house. They greet one another with smiles and then DJ asks Robert, " Do yo want to smoke?" Rob said yeah. DJ had some RoyalRolls (hand crafted preroll with THC glue) like in older times our ancestors would partake in the peace pipe and converse over commercial activity. So, DJ proceeded and got right to business. Explaining to Robert what his vision was for the company and that he even made a company name to do more marketing called GROWINGPHARMS. Which Sims at a previous date has told Robert he made Growing Pharms. DJ thought to himself "What a fuck boy" DJ had filed for the company name in Michigan so he had the documents to show that Sims was not being upright or forthcoming with Robert. They went outside continued to converse about hemp /cannabis and by the end of the speaking Robert had it in his mind that he would give DJ a Federal Exemption. He told DJ it would be a while till he got back to Jersey to handle the paperwork In DC. They shook hands and parted. About a month passed and Robert had another delivery in Utica Michigan about a hour away from Dj home. It was 1700 cubic feet 11900 pounds of weight that needed to be loaded on the truck and it was just Robert. So he called DJ to let him know he was coming in town again. DJ offered to help Rob with the Job. Rob said " Naw, you sure? A lot of people say they can do this kind of work." DJ laughed and said, "Yeah I got you." On the way there DJ asked if he could interview Rob and ask him some questions on video. Rob reluctantly agreed. DJ was on Facebook live and Robert didn't know until DJ ended the video. DJ expressed his gratitude and that live videos are the best because they catch genuine moment and they show truth. Shortly after they get to the job. It's an upstairs apartment with a wrap around side walk till you get to the truck. While Robert is talking to the client and wrapping pads on other pieces. DJ loaded the whole truck by the time Robert was finished with the documents. Robert looked in amazement and said, " You remind me of me" and laughed. So they headed back to DJ other house in the inner city Detroit to drop him off. Just conversing DJ expressed that his great aunt who was Ojibwe passed about a week ago. She would always talk about mustard seeds and how you should keep the faith of a mustard seed. DJ then reached in his pocket and pulled out a small fabric bag and went inside and pulled out some mustard seeds. Robert damn near lost his mind. He said " Wait, wait, wait you got mustard seeds?" Reason being back in 2007 Kenny Estrogens Roberts VP for music told Rob GAMPHACE is like a mustard seed." Something so small can feed many and it just grows and flourishes just like GAMPHACE." In that moment Robert looked down to the GPS to the next address for DJs home said 111 miles. 111 has held a special spiritual significance in Roberts life for years. With all these signs Robert said out loud, "hhhmmmm the Man with the Mustard Seeds". At this point Robert had decided the he was going to pass the torch. He gave DJ GAMPHACE and that changed everything.

THE TAKEOFF

From the Start in Oakland California to Patterson NJ to Detroit now with GAMPHACE in DJs hands, he contemplates what's next. Noto be corny or redundant but with great power comes great responsibility. Now at 27 years of age, hed just been given a multi-trillion dollar opportunity. With something no one else had and most wouldnt understand. What would you do? He did the best thing he could think of, MOVE TO MIAMI Florida. His reason for this is goes way beyond big booties and beaches. Florida is the only state that you can do outdoor grows year around. They have no existing state model there. So in his mind, a head start and no competition in the market and boy was he right.

So DJ planned a trip down to Florida in December for is birthate. He stayed about to week and did a spiritual retreat in Orlando. He wanted to be prepared for all that was to come with this commercial change. By the end of January DJ was right back in Florida holding his first company quarter meeting and giving a lecture at the same weekend for his other company A.T.U.M. in order to make it all make cents to him financially with GP not generating funds at this time. A month after, DJ packed up and just moved down to Florida with a suitcase full of clothes money and 2 cell phones all alone. Using some of the resources he acquired with his other businesses he had hotels discounted for 2 months he had 5 pounds of Meta-Sensation mailed down from Michigan Lawfully because of the exemption.

So between hotel hopping, still taking financial clients and expanding GAMPACE GROWINGPHARMS. DJ had meeting after meeting with all types of people. Being invited on to large estates with peacocks. Looking like some shit right out of a mob movie. The energy screamed federal indictment. DJ learned fast but took him a while to realize he himself had become a walking dollar sign and that's exactly how people started to treat him.

Using YouTube and his ATUMWORKS platform, he expresses the position he was in and he started to get request to not only do financial lectures but those on federal cannabis as well. From Georgia to Oregon then Kansas. That didn't include the other cities where people were just requesting for DJ to come out just to offer him a deal like Chicago and his hometown Detroit. From the time DJ moved to Miami he was on planes every week doing lectures and networking. DJs first contract he got in Florida that he accept the was a owner of a tow truck company. DJ knowing the rules to the game being brought up by , marines and school teachers OPM was the name of the game , Other Peoples Money. So he know he had to make the investor knowledgeable and comfortable or get someone whoes greed out weighed their knowledge. Just to get GAMPHACE on the map in Florida. DJ was brighter then most gave him credit for. Utilizing some of the members in the network before him. He was member 119 and had brought in others after himself. He called one member who was ex law enforcement and whose twin brother works in the Pentagon. DJ had the owner of the company, Tony profile pulled. Mainly because when DJ met with him it didn't sit well with him how much power this man had in this town. He lived just to be the owner of a tow truck company. Not to leave out Tony set up a meeting with DJ at his office. DJ drove up from Miami with his brother in the car. He walked in the office and to his surprise it's a off duty DEA agent and the Sherriff of the county along with Tony's adult children and his farmer making a total of 7 people in one room and DJ by himself. Tony said to DJ " I know you cant be on bullshit , if you were you wouldn't of showed up today." DJ just smiled and shook everyones hand. He had just got done smoking Starburts before he walk in so he wasn't worried bout shit but closing the deal after driving 2 hours with 4 pounds in the car. He greeted everyone then turned to the DEA agent and the sheriff who were sitting next to each other and asked what questions they had. DJ could tell by the body langue in the room that they were all buddies even Tony. They pulled out a folder with Roberts name on it and pictures from the original grow house in Jersey where he did 144 plants in the back yard and not ever had a problem. DJ asked, "What does that have to do with anything?" They Sherriff then said, "Well what about this time when he was arrested?" and DJ responded, "You mean when they dropped the chargers and ordered the return of federal property to him, you mean that time? " and they sat in silence. So he pulled out the company Bylaws which Florida Governor Ron DeSantis hold a copy in the State office and showed where it talks about the Rights of Indigenous people with him and Robert being tribal. The next section the gave the document to the DEA agent to read and it said I.R.S. Department of the Treasury registration removes members from a suspected class. C.S.A. 822(c)(1)(2https://www.deadiversion.usdoj.gov/21cfr/21usc/822.htm) allows for possession of controlled substance (MARIJUANA), warehouse and carriage. I.R.S. Department of the Treasury Federal I.D. #'s (20-********) allows the manufacture of a Schedule 1 controlled substance (Marijuana). Private business entity and business (members) are exempt from all statues of alleged criminal activity by the closed regulatory system enforced by the Drug Enforcement Agency (D.E.A). The Business has a closed distribution system that allows product (INDUSTRIALHEMP-CANNABIS-(META-CENSATION©™) for sale to members who are ultimate users; (there is no conveyance to a third party.) For the record. Business to apply a Sales of Tangible Personal Property for use directly and exclusively in experimental research and development in the laboratory sense. (tax). Research and development of [INDUSTRIAL-HEMP-CANNABISFLOWER] including "META-CENSATION©™" Carbon Sequestration.

DJ just looked at him as he read it and it looked like he wanted to break out his own skin. He flexed his neck like he was lifting weights in frustration. He handed DJ the document and leaned over in the room full of people and whispered "THAT SHIT DON'T MEAN NOTHING TO ME" and sat back up straight. This is a Field DEA agent that kicks doors and does raids. DJ thought to himself "IMA HAVE TO SUE THIS BITCH!" DJ then asked infront of the whole room aloud, " so are you say I cant do this?" They responsed no, just need to check some things out. DJ agreed, got his things , everyone shook hands and left immediately. His brother sat in the car the whole time just incase. When DJ got in the car Mill said,"I don't know what type of shit you're into bro but I seen like 20 cops in out that building while you was in there." As soon as DJ got down to Miami he switched cars. Call it what you want he this game and he would not sleep well until he got into another car.

Two days later DJ recieved a phone call from Tony and he informs DJ the DEA agent called him and wanted to apologize. He said "Everything that young man said was right it all checked out!" That's all Tony needed to hear, he was all in. DJ never gave a damn bout what they said he already knew what he had. He did that for the benefit of the company to have a known millionaire in florida. That people knew and let him open the market up after he fucks himself. DJ just knew Tony would always walk himself out of a deal before he could make any money off of him so he had to make sure word spread fast and Tony arrogant ass could not wait to run his mouth. This now put GAMPHACE GROWINGPHARM in the scope of multiple millionaires and a few billionaires. In DJ eyes, leverage is exposure, exposure is power and power is growth and it did just that. After meeting a plethora of fuck boys with money

DJ was introduced to a 20year restaurant owner Rocky. Rocky had I background in sales and a large cannabis deal the went wrong for the investors. The owner of the company ran off with everyone's money. So after coming out a burnt deal he did established his own brand, American Green Rose. The only FDA approved and pharmacist formulated brand. With all the changes in the market with the FDA standards on human consumption, Rocky was sitting on a gold bubble waiting to pop. So after a 5 hour meeting at a local Starbucks. DJ and Rocky agreed they could benefit each other. A few months later Rocky came down to DJs house in Miami Beach. DJ requested that Rocky become the director of Global Expansion for the company and he accepted. Due to his resources the company has expanded all over South Florida in just months and launched multiple CBD and THC brands under private labels.

While Rocky was working in Florida DJ was jumping on planes and this time to Atlanta and through his networking got a MOU (Memorandum of Understanding) with the Government of Belize to educated them on cannabis and bring financial stability through natural commodities. Which was not only major for the company but for cannabis. These thing happening in the background and the rest of the world has no knowledge till now.

November 24 after throwing a Federal Halloween Cannabis party DJ was introduced to Chief Tecumseh Brown Eagle of the Erie Indian Moundbuilders Tribal Nation because of DJ tribal status and the Federal Exemption, DJ was offered a position as the business director over the tribe and in exchange offered the Tribe the Exemption. Chief and DJ AKA Red Horse met in Erie PA at a library on a mound and signed the MOU for the Tribe and GAMPHACE GROWINGPHARMS . By January , DJ released a text book on the Erie Indian Moundbuilders tribal nation history. Also Put himself to be next Chief of the tribe due to all of his works.

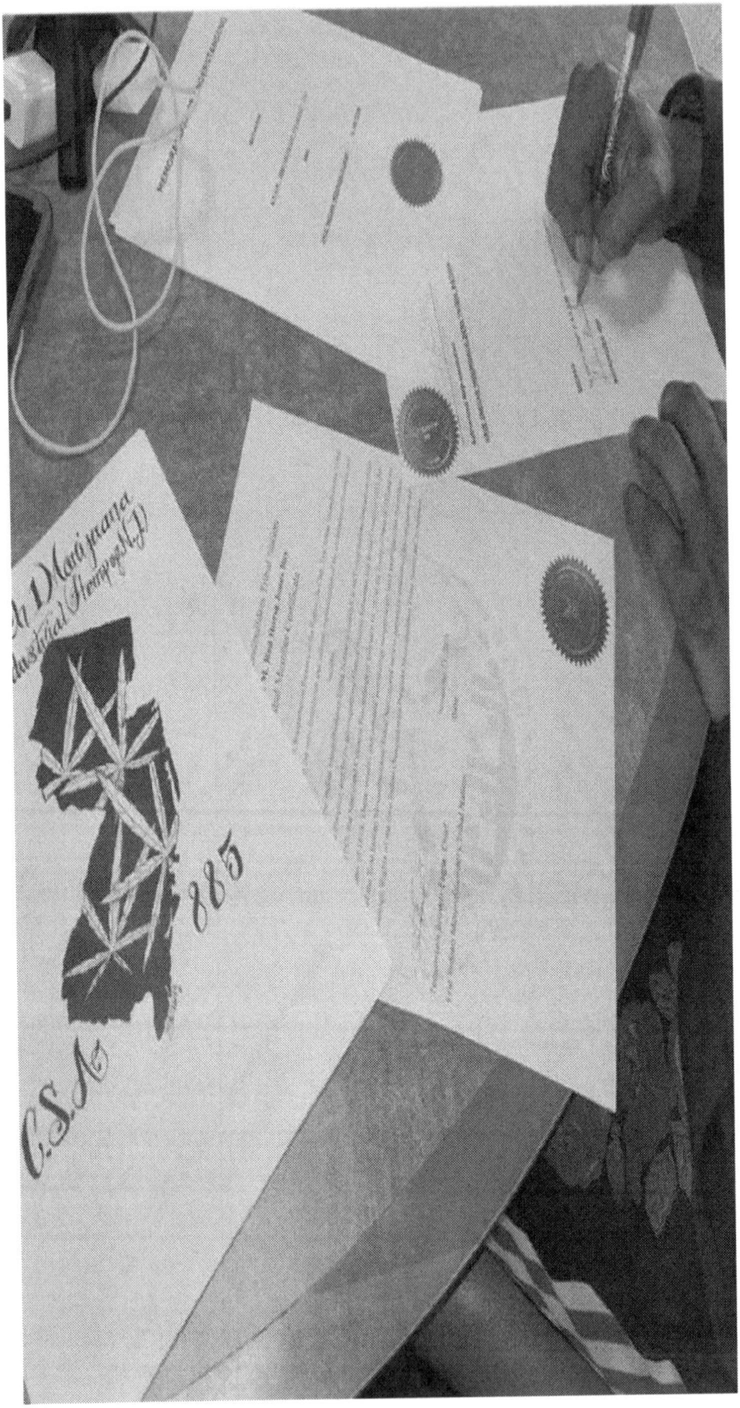
C.S.A
885

Now in 2020 DJ has brought back the Entertainment part of the company making GAMPACE ENT. Music being a universal language it directly effects peoples behavior (Doctrine of Ethos), its just like cannabis and its also how all this started. So with the market growing and the other company resources DJ started to acquire young artist and writers to teach them the business part of the game. Allowing them to be knowledgeable and able to acknowledge a bad contract while showing them how to fund their own projects so they could go to the labels with money.

Another by-product of legal hemp will be tens of thousands of new jobs across multiple sectors in the very near future. Besides hiring workers in agriculture, processing and manufacturing, the still-budding industry — with $1.1 billion in revenues 2018, estimated to more than double by 2022 to $2.6 billion, according to New Frontier Data — will need accountants, lawyers, compliance officers, government regulators, IT specialists, financial and insurance experts, transporters, researchers and lab technicians, marketers, CFOs, CEOs and various retail employees.

Since 2016, Vote Hemp has produced an annual Hemp Crop Report documenting the licensed cultivation of hemp in the United States. The annual hemp cultivation data and other statistics are provided by state departments of agriculture. Where states were unable to provide data on the hemp acreage planted, we estimated at 70% of licensed hemp acreage. This estimate is based on data collected in other states that has shown an average of 70% planting rate.

2019 U.S. Hemp License Report – 511,442 acres licensed
A full report listing licensed acreage by state. That's not even touching the THC market which GAMPHACE having the ability to deal in all variations of the plant. Puts the company at a MULTI TRILLION COMPANY VALUATION. For over 15 years people have

been lied to for the benefit of wealthy people in control they purposely slow down the hemp cannabis market production and laws to better align them selves financially. Can you blame them?

Like Robert always say "THE HATERS GOIN COME KEEP YOUR GAMPHACE ON!"

So either cry about it or make a change.

Its up to you.Moor to come !!!

........................Literally

10
National Currency
UNITED STATES OF AMERICA
R368589H
10
THE
SECOND
NATIONAL BANK OF
PATERSON
810
810
TEN DOLLARS
10
23057

A.T.U.M WORKS™
ALLEGIANCE TO UNITE MINDS
PUT YOUR
GAMPHACE™
ON!!
G -Game Plan
A -Attitude
M -Motivation
P -Pride
H -Hunger
A -Ambition
C -Confidence
E =Excellence

Made in United States
Orlando, FL
05 November 2026

90230230R00025